Contents

EARLY METHODS OF TELECOMMUNICATION

The word **telecommunication** means the transfer of information over large distances. Nowadays this tends to conjure up visions of computers, satellites, high technology and other scientific mysteries beyond the understanding of ordinary folk. But many of the basic ideas of telecommunication have been understood for centuries. Looking at early methods of telecommunication shows how modern techniques have developed. Further advances are still being made and will continue.

Noisy communication

Some early methods of telecommunication used **sound energy**. Perhaps the simplest of these is shouting. Shouted messages do not travel very far. The range may be improved slightly by shouting into a large cone held close to the mouth. This directs more sound energy towards the listener. By spacing people out within shouting distance of each other, messages can be sent fairly quickly from one place to another. But shouted information is completely public. Everyone within earshot hears the message.

Drums also use sound energy to transfer information. The sound of drumbeats travels much further than shouts, in some cases up to 10 kilometres! Many people can hear drums but not everyone understands the code. Unfortunately using a drumbeat code may limit the number of messages which can be sent.

This drummer is a king's bodyguard in Kenya

By foot or wing

Other early methods of telecommunication used **people** or **animals**. Human messengers could carry written or memorized information anywhere it was possible to travel. This information moved very slowly from place to place as most early messengers had to walk or ride. Information carried by messengers *could* be completely private, though it was not always. Messengers could also carry replies, providing two-way communication.

Pigeons were probably the most common animal messengers. Though much faster than humans, they could carry only short messages in only one direction.

Messengers are still used nowadays but most people say this is not real telecommunication.

A pony express rider, 1861

Messages by light

Some early telecommunication methods used **light**, for example polished sword blades reflecting sunlight, smoke signals and fires. Some of the important features of modern telecommunication can be checked using light from torches. For example, if you want to send messages

a light (energy) must be able to get from the sender to the receiver
b the receiver must know where to look
c the receiver must know when to look
d the receiver should be able to send replies
e both the sender and the receiver need to understand the code.

Perhaps you could try out some of these ideas yourself. Try to use torches to send a simple message across a room. With a little practice you might be able to use torches to transmit messages in both directions at the same time (though attempting this is likely to make your teacher cross-eyed!).

FASCINATING FACT!

Pheidippides, a messenger in ancient Greece, was the inspiration for the modern marathon.

NOW TRY THESE!

1 What does **telecommunication** mean?
2 Estimate the maximum distance *you* can shout a message. Check your estimate by experiment.
3 Why is it difficult for people to shout messages to each other on windy days?
4 Either by experiment or by making some intelligent guesses estimate the number of people needed to send a shouted message 2000 metres.
5 Give one modern example of someone using a cone shape to make shouting more effective.
6 Write down four important ideas in telecommunication.
7 **a** What do you think a drumbeat code is?
b Listen to some of your favourite records. Try to identify different drum rhythms. Write down the number you can identify and compare your number with other members of your class.
c Do you think that using a drumbeat code limits the number of messages which can be sent?
8 Explain how you could use a torch to send a message over a large distance.
9 Find out about the **semaphore code**. Make a poster or write a one-page report.
10 Try to decide which of the early methods of telecommunication is the best.
Give *each* method a score (out of ten) on *each* of the following:
a **Speed** – how quickly do the messages travel?
b **Range** – how far can the messages travel?
c **Privacy** – how many people can receive the message? (large number of people, low score, etc.)
d **Reliability** – how confident can the sender be that the message will get to the receiver?
e **Content** – can lots of different messages be sent?

Copy the following table into your workbook and enter all the scores you give.

method	speed	range	privacy	reliability	content
shouting					
drums					
pigeon					
mirror					
smoke signal					
signal fire					

HOW QUICKLY DO MESSAGES TRAVEL?

Some methods of telecommunication are faster than others. For example, in AD 1805, news of the Battle of Trafalgar, carried by messengers, took 16 days to travel the 2000 kilometres from the south of Spain to London. Yet today, when any British team plays an important football match in Spain, you can watch the game 'live' on television (as soon as you finish your homework!).

The speed of a message depends on the form of energy which carries the message.

Measuring speed

On way of finding speed is from measurements of **distance** and **time**. The measured distance is divided by the measured time. Speed is usually measured in **metres per second**, though it can be measured in other units such as **m.p.h.** or **km/h**.

This speedometer shows m.p.h. and km/h

Sounding it out

Experiments to measure the speed of sound used to use stopwatches and needed a lot of space. Nowadays we can measure the speed of sound inside a laboratory. For these experiments you need something to make a loud noise, three *micros* (two *micro*phones and one *micro*computer) and a timing program for the computer. The microcomputer is used because it is the most accurate time measuring device in the school. A microcomputer can measure time to within one millionth of a second. If your physics teacher is very clever she or he will set up this experiment for you! The apparatus should look like this.

THE EASY EXPERIMENT

The first microphone is connected to *start* timing. The second microphone is connected to *stop* timing. Let your teacher worry about the technical details! Load the timing program into the computer.

Measure the distance between the microphones.

Bash the hammer on the metal block. The microcomputer automatically measures the time for the sound to travel between the microphones. The measured time is displayed on the screen.

Divide the measured distance by the measured time.

Sample results:

distance between microphones = 2.00 m
time measured by microcomputer = 0.006 s
speed of sound = 333 m/s

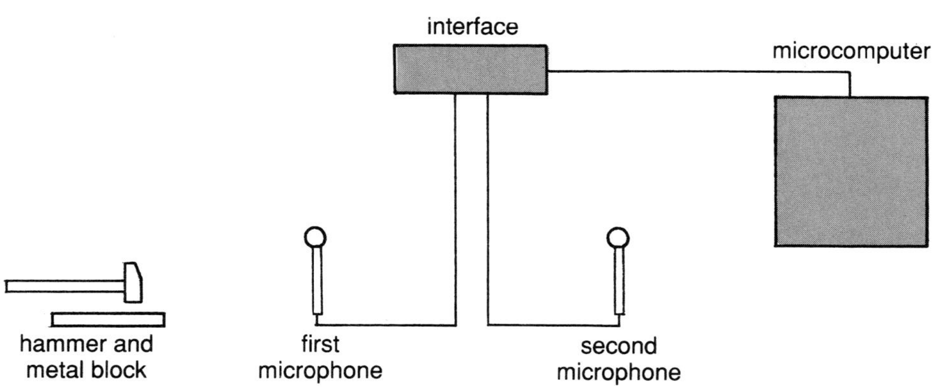

THE EVEN EASIER EXPERIMENT

Start in the same way as before. When the timing program is loaded add a few lines to make it into a speed calculating program. You have to tell the computer the distance between the microphones. Then tell it how to work out the speed. Perhaps your timing program already has a speed calculating part.

Bash the hammer on the metal block. By the time you look at the screen the computer has worked out the speed of sound. This is so quick that everyone in the class could take a turn. You could even ask the computer to work out the average if you know how to change the program.

FASCINATING FACT!

Light is *slower* in glass than in air.
But sound is *faster* in glass than in air.

Quick as a flash

Light is about a million times faster than sound. Electrical signals in wires travel just as fast. Experiments to measure these speeds are difficult to explain. So let's skip the experiments and get to the results. After all you know you can trust your physics teacher!

The speed of light in air is nearly 300 million metres per second. Electrical signals in wires travel at this speed too!

NOW TRY THESE!

c

1 Using the information in the text, work out the average speed of the news of the Battle of Trafalgar on its journey from Spain to London. Give your answer in m/s.

2 What do m.p.h. and km/h mean?

3 What is a computer program?

4 Why is it better to repeat an experiment several times than to do the experiment only once?

5 Why do you think older methods of measuring the speed of sound needed lots of space?

6 Give another name for a time measuring device.

7 How long does it take sound to travel 10 km in air?

e

8 How far would light travel through air in 1 second?

9 What is an echo? Try to think of a way of using an echo to measure the speed of sound.

10 Try to write the few extra lines which would be needed to change a time measuring program into a speed calculating program. Write the extra lines in BASIC.

11 Find out about a **light year**.

12 a How long does it take light to travel 10 km?
   b How far would sound travel in this time?

13 a Work out the speed of sound in km/h.
   b Work out the speed of light in km/h.
   c Work out the speed of sound in m.p.h.
   d Work out the speed of light in m.p.h.

Signals in Wires

DAILY BLURB

4 May 1837

Scientific Scoop

by Steel A. Story

Modern science continues to amaze. World leaders in telecommunication, Cooke and Wheatstone, yesterday demonstrated their latest success to officers of the main railway companies.

These great scientists actually managed to send a message from one room to another...*through wires*! No one spoke! No flashing lights! No movement of the wires! Nothing!

The passing message is completely silent. There is no glow. The wire does not get hotter or colder. In fact, there is no way of telling when there is a signal in the wire. And there is no way of telling what the message in the wire is. Messages in these wires are *totally secret*.

This was definitely not a trick. This reporter checked all of the equipment before, during and after the demonstration. I was even allowed to operate the machine.

The machine is called a **telegraph**.

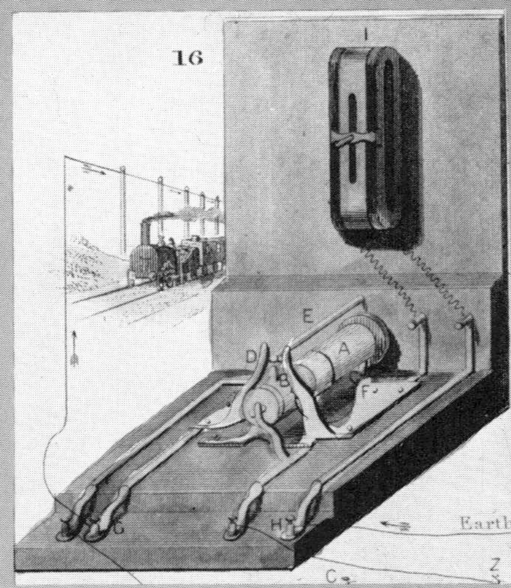

Cooke and Wheatstone's telegraph

Cooke explained how the equipment could be used to make the railways safer. When a train leaves one station, the station-master could send a message to the next station. Then the people there would know a train was coming along before they could see it or hear it. The men from the railways seemed to be very impressed.

Wheatstone said that the wires could carry messages over very large distances. All you need is a very long wire. Also the messages travel very quickly. He said "in the twinkling of an eye".

This machine could replace letter writing. Why write letters when you can communicate instantly through wires? Every home should have one.

Why were telegraph wires put alongside railways?

Improvements

As you can see from the date on this made-up newspaper story, the idea of the telegraph has been around for a very long time indeed. And Cooke and Wheatstone were not the first to think of it.

The early telegraphs could carry only one message at a time. But by the 1860s it was possible to send messages in both directions along the same wire at the same time. This was called **duplexing**.

Later, Frenchman Emile Baudot found a way of letting several people use the same line at the same time. His method is called **time division multiplexing**. It is a bit like taking turns. But everyone gets a turn every fraction of a second. Time division multiplexing (TDM) is still very important in modern telecommunication systems.

An international telegraph network developed in Europe. Later the idea was taken to America by an artist called Samuel Morse. He also invented a telegraph code which was used all over the world. This code is still well known today. The modern **telex** system developed from the telegraph and has become a world-wide network.

Did you notice that the reporter has made some mistakes in the scientific 'facts' he uses?

A modern telex machine

NOW TRY THESE!

c

1 The reporter clearly thought the telegraph was wonderful. Why was he so impressed?

2 Compare the telegraph with sending messages by drums

- **a** In what ways is the telegraph better?
- **b** In what ways are the drums better?

3 **a** Find out the Morse code for SOS.

- **b** What does SOS stand for?
- **c** Why is this such an important message?

4 Telegraph messages travel at nearly 300 million m/s. How long would it take for a telegraph signal to travel 40 000 km? (40 000 km is about the length of the equator!)

5 What is a duplex operation?

6 Cooke and Wheatstone wanted to sell their invention to the railways. Why do you think they chose the railways?

7 Steel A. Story made scientific mistakes in his report. Do modern reporters make scientific mistakes?

e

8 The reporter said that there was no way of knowing when wires were carrying a signal. He was wrong! Can you think of a way?

9 How can telegraph messages be stopped?

10 Find out about the Morse code. Write a one-page report.

11 Try to send a simple message using the Morse code. You can make it more fun by making it into a competition. Make up a sentence of around ten words. Find out who can send this message most quickly! Your teacher *might* give the winners a prize.

12 The story only mentions the 'men from the railways'. Why do you think there were no women officers from these companies?

13 Which modern telecommunications system(s) use TDM?

A Life on the Ocean...?

Telecommunication is possible only if **energy** can travel from one place to another. The energy carries the information we want to send. But

- sound signals travel in waves
- light signals travel in waves
- electrical signals travel in waves
- radio signals travel in waves!

In fact all of the forms of energy used in modern telecommunications travel in waves. Understanding a little about waves will help you understand telecommunication better.

What are waves like?

If you have ever had a bath you might have seen little ripples on the surface of the water. That's what waves are like! Waves carry energy, not materials, from place to place. So when you make ripples on the surface of your bath the water doesn't all go down to the end of the bath.

Here are some words used to describe waves:

amplitude is the height of the waves at the highest bit or the depth of the waves at the lowest bit. The bigger waves are, the more energy they carry.

frequency is the number of waves made in 1 second. The frequency of waves is measured in hertz (Hz). Hertz is the same as 'waves per second'.

speed is how fast the waves travel. It is just like any other speed. But waves change speed only when they go from one material to another.

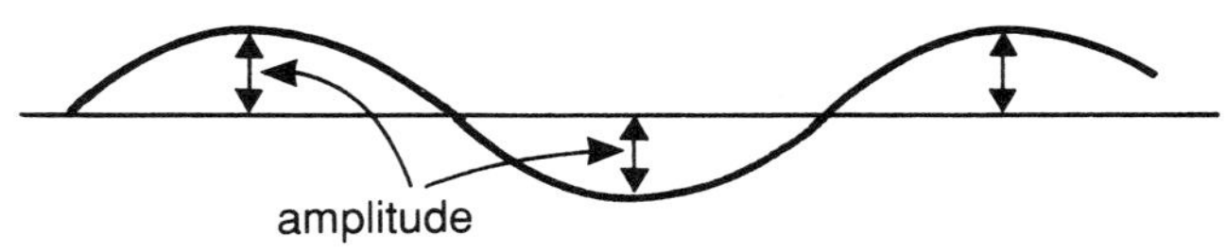

Amplitude is height or depth

What can waves do?

You can do all sorts of things. But waves can do only certain things.

Bouncing

The waves can bounce off objects which get in their way. This is called **reflection**. Waves bouncing against objects can sometimes make telecommunication more difficult. Sometimes it can help. The reflection of waves is very important for satellite communication.

The reflectors on a telecommunication satellite

Faster, slower

Also, waves can move from one material to another. This is called **refraction**. Usually waves travel at different speeds in different materials. So when waves refract, their speed changes. Refraction is very important for telecommunication using optical fibres.

Some materials let only some kinds of wave travel. For example, sound cannot travel through a vacuum but radio can. Electricity can travel easily through a wire. But have you ever tried shining a light through a wire?

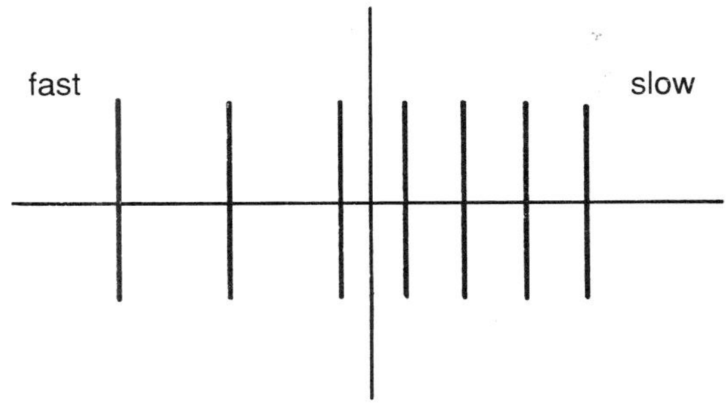

Refraction – the speed changes as the waves pass from one material to another

Round the corner

Sometimes signals have to be sent over mountains and into valleys. This is only possible because waves can actually bend around corners. This is called **diffraction**. Diffraction is very important for radio and television.

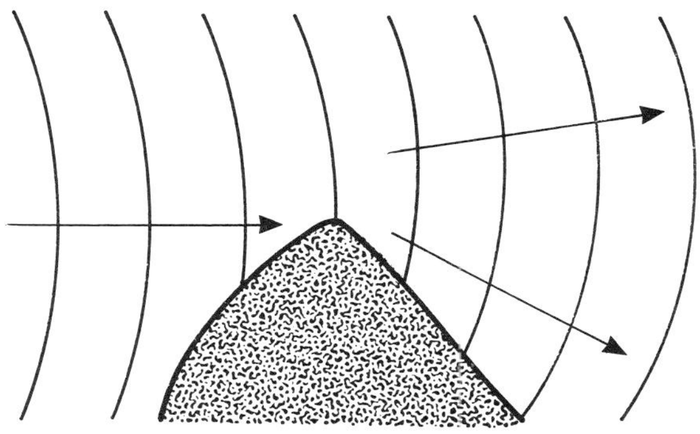

Diffraction over a mountain

The words reflection, refraction and diffraction are easy to get mixed up. You can save yourself a lot of trouble by learning them properly now!

Telecommunications scientists have to choose different forms of energy to do different jobs. It all depends on the kind of waves and how the waves behave (or misbehave). When you are learning about telecommunication, try asking yourself why the scientists have chosen certain kinds of wave.

c

1 Why are waves important in telecommunications?

2 What happens when waves are reflected?

3 Does the speed of a wave change when it is reflected?

4 Write down two early methods of telecommunication which used reflection.

5 What happens when waves are refracted?

6 Name one piece of telecommunication equipment which uses refraction.

7 What happens when waves are diffracted?

8 Does the speed of a wave change when it is diffracted?

9 Use the idea of diffraction to explain why you can still hear your mother talking to you when she is in the kitchen but you are out in the garden. Remember sound energy travels in waves. (Saying your mum shouts a lot is not the correct answer!)

10 Sound waves travel faster in iron than they do in air. Does this mean that sound waves slow down when they come out of iron into air?

NOW TRY THESE!

e

11 Light travels at around 300 million m/s in air. It can only manage around 200 million m/s in glass. Work out the time for light to travel 150 million kilometres in air.
– now work out the time for light to travel 150 million kilometres in glass.
– (The sun is roughly 150 million kilometres away.)

12 Find out about **ghosting** of TV pictures. Explain why reflection of TV waves might cause ghosting.

13 Why might it be difficult for miners to listen to a radio at work?

talking through wires

Hard luck Elisha

On 14 February 1876, Elisha Gray went to the patent office to register his invention – the **telephone**. He was too late! Alexander Graham Bell had beaten him to it. Bell had registered his telephone just a few hours earlier. Both men had the same idea at the same time. Bell has become world famous because of those few hours. Hardly anyone knows about Gray.

Who needs a code?

Gray and Bell had the same good idea. This idea is really quite simple. People speaking make **sound energy**. But telegraph wires do not carry sound energy. Telegraph wires carry **electrical energy.** Both Bell and Gray tried changing the sound energy into electrical energy. This electrical energy is sent through the wire. At the end of the wire the electrical energy is changed back into sound. This system does not need a code. The sound at the end is just like the sound at the start.

Changing energy

The telephone needs two energy changers. Energy changers are called **transducers**. So a telephone has two transducers.

One transducer changes sound energy into electrical energy. This is called a **microphone**. A microphone is an **input transducer**. It puts energy *into* the system.

The other transducer changes electrical energy into sound energy. This is called a **speaker**. A speaker is an **output transducer**. It puts energy *out of* the system.

The first telephones used the same piece of equipment at both ends of the wire. It was supposed to be a speaker and microphone rolled into one. It did not work very well. Early telephones were also very expensive.

A modern telephone – the speaker and microphone are both in the handset

An early telephone – the speaker and microphone are the same

But the idea was good and is still used in modern telephones. In a modern telephone the microphone is inside the mouthpiece. The speaker is inside the earpiece.

Electricity to sound

A telephone earpiece contains two kinds of magnet, a **permanent magnet** and an **electromagnet**. Permanent magnets are usually made of iron. A permanent magnet is magnetic all the time. Electromagnets can be made from any kind of wire but they usually have a piece of iron in the middle. An electromagnet is magnetic only when there is a current in the wire.

The diaphragm is attracted by the permanent magnet. When a current flows in the coil of the electromagnet the diaphragm is attracted even more. As the current changes, the diaphragm moves backwards and forwards. This makes the air vibrate. We call this sound energy.

The pattern of current changes produces a pattern of sound vibrations.

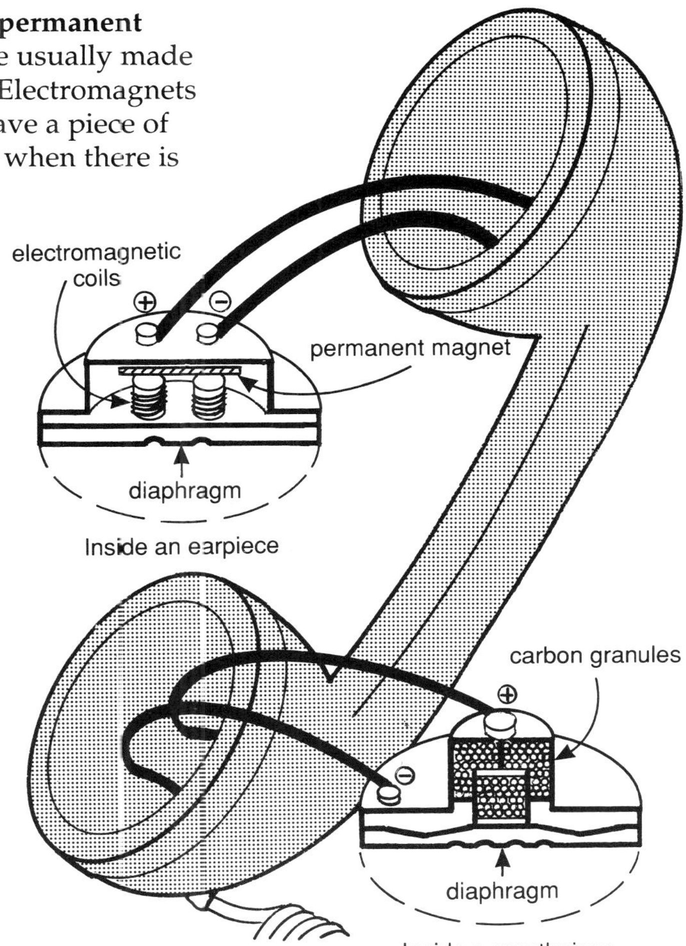

Inside an earpiece

Inside a mouthpiece

Sound to electricity

When someone speaks into the mouthpiece of a telephone, the sound energy makes the **diaphragm** move backwards and forwards. This squeezes and stretches the **carbon granules**. When the carbon is squeezed, a bigger current flows. When the carbon is stretched, a smaller current flows.

The pattern of sound vibrations produces a pattern of current changes.

Now try these!

1 Find out the meaning of the word telephone. You will need to use a special dictionary called an etymological dictionary. (You may have to go to a library but the librarian will help you.)

2 What is a patent office?

3 Why does the telephone not need a code?

4 What is a transducer?

5 Write down the energy change in a microphone.

6 **a** Write down the energy change in a speaker.
b Write down the names of two other transducers which have the same energy change as a speaker.

7 Explain the difference between an input and an output transducer.

8 Both the telephone earpiece and mouthpiece contain a diaphragm. What is a diaphragm?

9 **a** Why does the telephone microphone contain carbon?
b What are granules?
c Explain why a bigger current flows when the carbon granules are squeezed.

10 Explain the difference between a permanent magnet and an electromagnet.

11 Why does a telephone earpiece need both a permanent magnet and an electromagnet?

12 **a** Are ordinary domestic telephones electrically live?
b Find out the voltage used for ordinary telephones.
c Could you get an electric shock from a telephone?

Silent Shouting

Communication without wires

Even though the telephone was a great advance from the telegraph, it still suffered from one great problem – wire! All messages had to be sent through wire. This meant that very long pieces of wire had to be stretched all over the country. Also if the wire breaks, the telephone is useless.

Some scientists dreamed of communicating over very long distances without wire. They knew they would have to use a different form of energy, but no-one knew what to use.

In 1887, the German scientist Heinrich Hertz demonstrated something called **radio waves**. At first no-one realized how useful these waves would be. But ten years later, an Italian scientist Guglielmo Marconi was able to use these waves. He sent simple messages over short distances. He called his equipment the **wireless telegraph**. This was the first radio. Some older people still call a radio a wireless.

How fast?

In air, radio waves travel at the same speed as light waves. Radio waves carry messages almost instantly from place to place.

How far?

On 12 December 1901, Marconi sent a message across the Atlantic ocean. The message travelled a total of 3440 km. This is almost one-tenth of the earth's circumference. Nowadays radio messages are sent many times further than this.

Guglielmo Marconi and his wireless telegraph

What is a radio wave?

Radio waves are electromagnetic waves. They cause both electrical and magnetic forces. Also they have very high frequencies. There are lots of waves every second.

There are many different kinds of radio wave. So they are split into different **wavebands** like **HF** and **LF**. HF is short for **high frequency**. HF radio waves have frequencies around 30 million hertz. LF is short for **low frequency**. LF radio waves have frequencies around 300 thousand hertz.

These different kinds of radio wave are actually very similar. They are different because they have different frequencies. Also, they are used for different kinds of radio system. For example, taxis use a very high frequency of radio wave. But many radio stations use medium frequency waves.

A radio tuning panel shows different wavebands

Perfect telecommunication?

Radio was quite a discovery, but these magical waves are not perfect. Radio waves can travel very easily through air or even outer space. But they cannot travel through metals. If you put your radio inside a metal box it will not work. Try it and see! This sometimes makes it very difficult to send messages by radio.

Also radio signals can be affected by electrical storms. If there is a lot of lightning about you might not be able to listen to your radio.

FASCINATING FACT!

In 1864, James Clerk Maxwell, a Scottish physicist, proved that radio waves existed. No-one believed him!

James Clerk Maxwell

Now try these!

c

1 Why do you think this unit is called Silent Shouting?

2 Do you think telecommunications scientists were completely happy with the telephone? Explain your answer.

3 In what ways is radio better than the telegraph?

4 In what ways is radio better than the telephone?

5 Radio waves are electromagnetic. What does this mean?

6 In what year did Marconi send the first radio message?

7 Radio waves travel through the air at the same speed as light. What is the speed of light in air?

8 What does circumference mean?

9 **a** Work out how long it took Marconi's message to travel across the Atlantic in 1901.
b How long would it take nowadays?

10 Why do some people call a radio a wireless?

11 Is radio the perfect telecommunication system? Explain your answer.

e

12 **a** Write down the equation which links the wavelength, frequency and speed of radio waves.
b Use this equation to calculate the frequency of a radio station which transmits radio waves of wavelength 261 m.
c Write down the name of a radio station which transmits this frequency of radio waves.

13 What do the following labels for radio wavebands stand for?
a VHF
b UHF
c MF

14 Find out about at least three different uses of radio waves. Here is a list to help you choose, though you do not need to stick to this list.

VHF, UHF, CB, Radio Taxis, radio-controlled models, Coastguard

Write a one-page note on each waveband. Your notes should include
a the frequency and wavelength of the radio waves
b the range of a typical transmitter
c the cost of a transmitter and receiver
d how clear the messages are.
Compare your notes with other people's in your class.

15 Why does a car radio need an aerial outside the car while a radio in a house doesn't need an aerial outside the house?

Messages by Radio

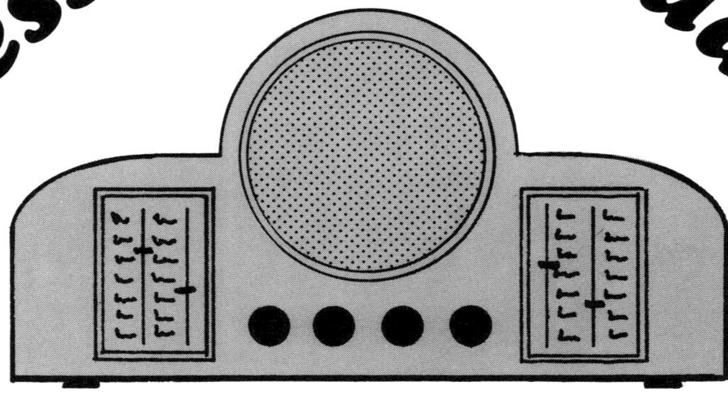

Here are the parts of a radio transmitter. This is the equipment used to send messages using radio waves.

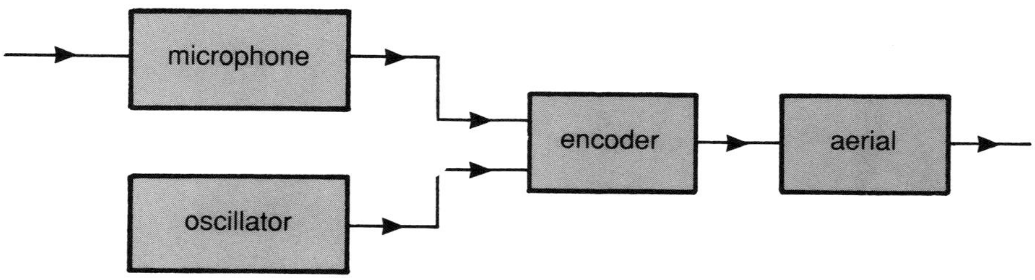

Radio waves have very high frequencies – sometimes several million waves every second! On the other hand, the sound waves which human beings can hear have quite low frequencies, only up to about 20 000 waves each second.

When a sound is made near a microphone the sound waves are turned to electrical waves. These electrical waves have the same low frequency as the sound.

A special kind of electrical circuit called an **oscillator** makes high frequency electrical waves.

The low frequency and high frequency electrical waves are added together. This makes the height of the high frequency wave change. The message is now carried by the changing height of the high frequency electrical wave. As this is like putting the message into code, the piece of equipment used to add the two waves together is called an **encoder**.

The combined electrical wave is sent to a **transmitting aerial**. At the aerial the electrical waves are changed to radio waves. The radio waves spread out in all directions. The height of the radio waves changes in the same way as the height of the electrical waves. This changing height of the radio wave is what carries the message away from the transmitter.

Another word for change is **modulation**. And another word for the height of a wave is **amplitude**. So this kind of radio transmitter is an **amplitude modulation** (AM) transmitter. It sends out AM radio waves.

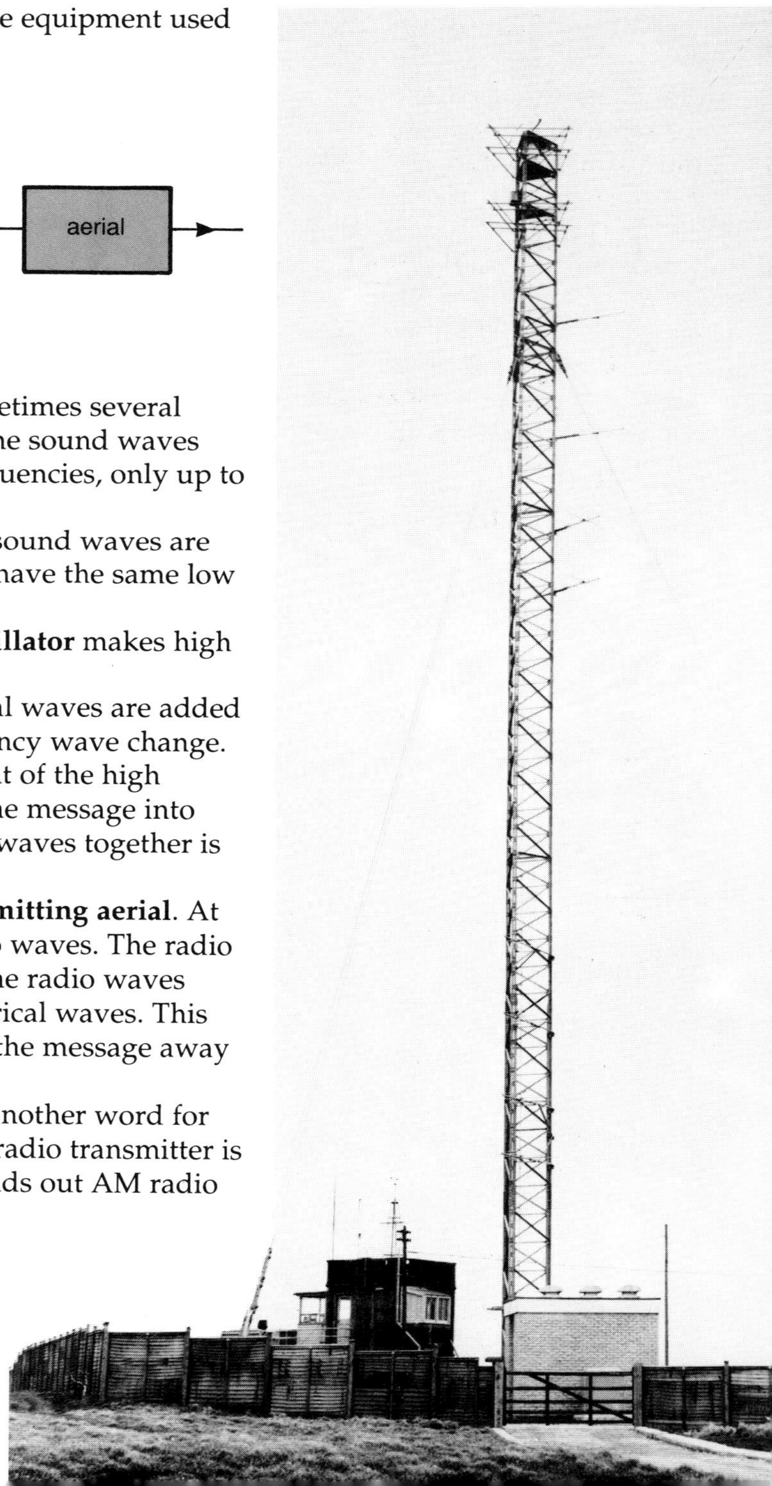

A radio transmission aerial

Getting the message

A **radio receiver** is a piece of equipment which can be used to pick up radio waves. Most people just call it a radio.

Look for the letters 'AM' on a modern radio. If you can find them it means that this radio is an AM radio. An AM radio can pick up AM radio waves. Some radios are marked 'AM/FM'.

Here are the parts of an AM radio.

At the aerial, radio waves are turned to electrical waves. These electrical waves are very high frequency, just like the radio waves. They are also modulated in the same way as the radio waves.

In this country there are lots of radio stations sending out radio waves. The aerial on a radio picks up all the radio waves which are going by so there are lots of electrical waves produced in the aerial. These waves have different frequencies and are all mixed together. So the electrical waves are sent to a special circuit called a **tuner**. The tuner picks out one radio frequency.

These waves then go to a special circuit which separates the high and the low frequency parts. As this is like taking the message out of code, this circuit is called a **decoder**.

Do you remember that the low frequency part of the wave was produced at a microphone? The low frequency bit then goes to a speaker where the electrical energy is turned back to sound. The sound produced at the speaker is just like the sound made at the microphone.

Free radio?

A very simple radio circuit does not need to be connected to the mains supply or to a battery. It works without them. All you do is build the circuit and you can listen forever for free!

Simple radios are usually very quiet. You can listen only on a small earphone. So most radios are more complicated. They have another special electrical circuit called an **amplifier**. An amplifier makes the electrical waves bigger before they are sent to the speaker. This makes the sound produced at the speaker much louder.

The amplifier circuit needs a supply of electrical energy. Radios use batteries or mains electricity to provide this energy.

Now try these!

C

1 Why are radio stations called transmitters?

2 How are radio waves different from sound waves?

3 Can people hear radio waves?

4 **a** In radio, what does AM mean?
b What do you think FM means?

5 What happens at the encoder?

6 Write down the energy change at a radio transmitter aerial.

7 What is the name given to the kind of electrical circuit used to make high frequency electrical waves?

8 Why is a radio wave called a carrier wave?

9 Write down the energy change at a receiver aerial.

10 What happens at the decoder circuit?

11 Why does a radio need a tuner circuit?

12 Why do radios need electrical energy?

e

13 Find out about FM radio transmission. Try to find the answers to questions like
- What are the problems with AM transmission?
- Is FM transmission better?
- How does FM transmission work?

Write a one-page report or teach the rest of the class about it.

14 It is possible to build a radio which does not need a battery or mains.
a Find out how to build a radio like this.
b Build one and get it working.
c Where does its energy come from?

TELEVISION

John Logie Baird

Early this century there was a race to be the first to send pictures using radio waves. Perhaps the most famous competitor was the Scot John Logie Baird. He won the race with his mechanical system. But another system was developed by Russian Vladimir Zworykin. Hardly anyone has heard of him. Zworykin's system was electronic.

At first the BBC used both systems. One system was used one week and the other system was used the next. In 1936, at a special exhibition called Radiolympia, there was a competition between the mechanical and the electronic systems. The electronic system won.

Some people say that nowadays, the mechanical system would win! But modern televisions still use the electronic system.

John Logie Baird's mechanical system

Electronic television

Here is an outline of a television transmission system.

The system is a bit complicated so think about each little bit on its own. Then try to understand the complete system.

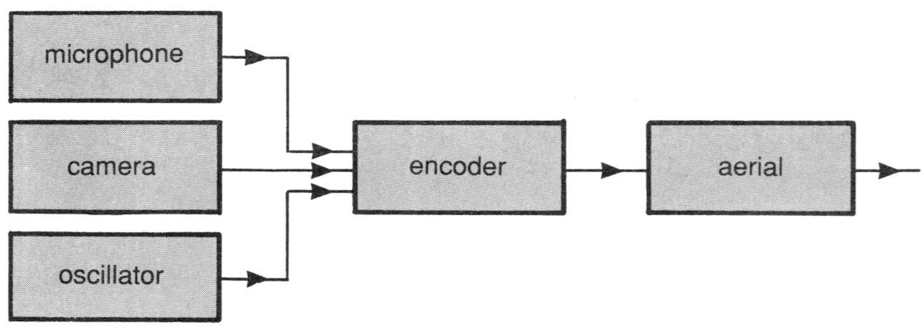

1 A sound wave is turned into a low frequency electrical wave by a **microphone**.
2 At the same time light is turned into a high frequency electrical wave by the television **camera**.
3 Also at the same time a special electrical circuit called an **oscillator** makes a very high frequency electrical wave.
4 At the **encoder** all three electrical waves are added together. This makes a modulated very high frequency electrical wave.

5 The modulated electrical wave goes to a transmitter **aerial**. Here the electrical wave is turned into a radio wave.
6 The VHF radio wave carries the message through the air to the **receiver system** shown here.

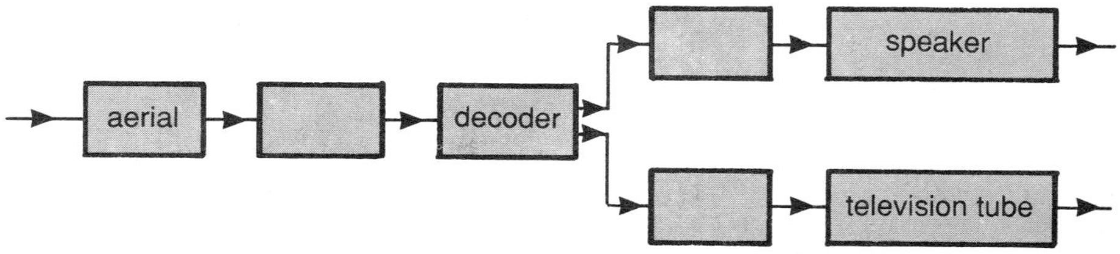

7 A television **aerial** picks up the signal. This signal goes to the television set's electronic circuits.
8 The television set's **electronic decoder circuits** separate the different electrical waves.
9 The low frequency electrical wave goes to the **speaker**. Its electrical energy is changed to sound energy. It becomes the sound from the television.
10 The high frequency electrical wave goes to the **television tube**. This electrical energy changes to light energy to make the picture.

Quite a system!

Fast!

Describing the television system takes a long time. Using a television system is much quicker. Altogether the ten stages take only a tiny fraction of a second! Live broadcasts really are live!

Vision of power

A television transmitter has to send out picture information and sound information. So it needs more energy than a radio transmitter. It has to be more powerful. This makes building a television station more expensive than building a radio station.

Filling the spaces

The television transmission system sends lots of pieces of information one after another. So there are small spaces. Nowadays the spaces are used to send extra information. This is called **teletext**. To get at this information you need a specially adapted television set.

In this country there are two different teletext systems, one on the BBC and the other on the IBA channels. Both systems work on British televisions.

When your television set is put into the teletext mode you can choose to read any page being transmitted by the television station. The information available covers a wide range of topics from the day's news to jokes and puzzles. You can even find out if it will rain tomorrow!

A weather page from Oracle

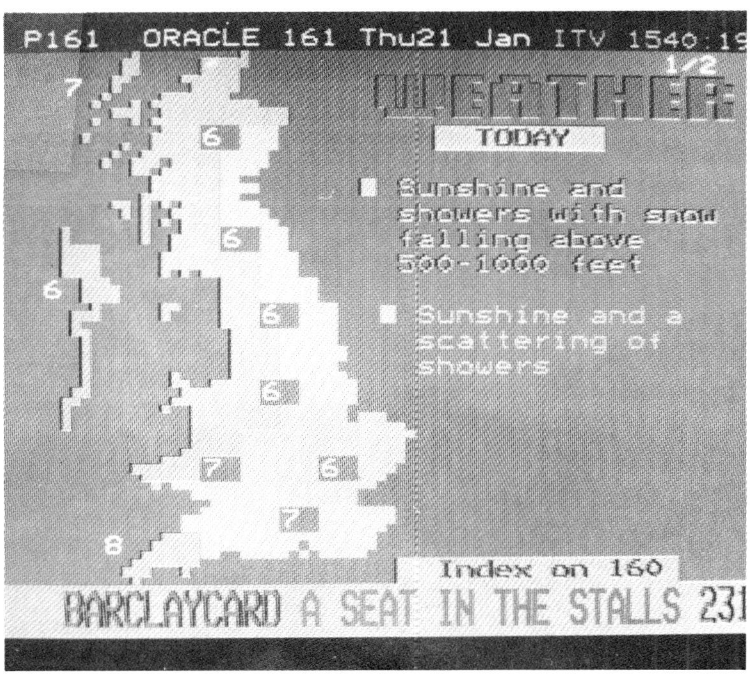

On a teletext television set you can get the normal television picture or the teletext screen on its own. But if you want you can mix them so that both are on at the same time.

Now try these!

c

1. **a** At first, how many television systems were there?
   b How were they different?
2. What system do modern televisions use?
3. **a** Why do you think that Logie Baird is famous but Zworykin is not?
   b Do you think Zworykin should be famous?
4. What does television transmission system mean?
5. Write down the energy change in a television camera.
6. What does the oscillator do?
7. What kind of waves are used to carry television signals?
8. Write down all the energy changes in a television set.
9. **a** How many electrical waves are added together at the encoder?
   b How are these waves different?
10. Why does a television transmitter send out more energy than a radio transmitter?

e

11. Do you think the television system is like the radio system?
    a Make a list of ways in which the two systems are similar.
    b Make a list of ways in which the two systems are different.
12. Find out how a television camera works. Write a one-page note including a diagram.
13. Find out how radar works. Why does radar need a good picture tube just like a television?
14. What do you think should be in the empty boxes in the diagram of a television receiver system? Explain your answer.

PUTTING US IN THE PICTURE

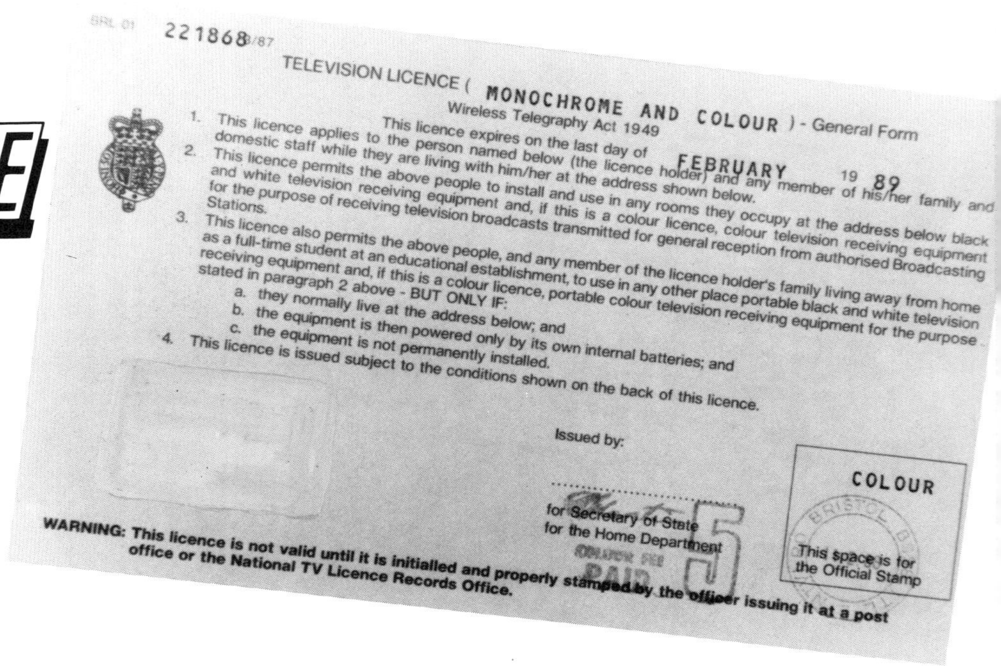

221868 3/87

TELEVISION LICENCE (MONOCHROME AND COLOUR) - General Form
Wireless Telegraphy Act 1949
This licence expires on the last day of FEBRUARY 19 89

1. This licence applies to the person named below (the licence holder) and any member of his/her family and domestic staff while they are living with him/her at the address shown below.
2. This licence permits the above people to install and use in any rooms they occupy at the address below black and white television receiving equipment and, if this is a colour licence, colour television receiving equipment for the purpose of receiving television broadcasts transmitted for general reception from authorised Broadcasting Stations.
3. This licence also permits the above people, and any member of the licence holder's family living away from home as a full-time student at an educational establishment, to use in any other place portable black and white television receiving equipment and, if this is a colour licence, portable colour television receiving equipment for the purpose stated in paragraph 2 above - BUT ONLY IF:
   a. they normally live at the address below; and
   b. the equipment is then powered only by its own internal batteries; and
   c. the equipment is not permanently installed.
4. This licence is issued subject to the conditions shown on the back of this licence.

Issued by:

for Secretary of State
for the Home Department

COLOUR

This space is for the Official Stamp

WARNING: This licence is not valid until it is initialled and properly stamped by the officer issuing it at a post office or the National TV Licence Records Office.

A colour television licence

In a television the picture is made in a glass tube. There are two kinds of television tube. One kind can produce only one colour, usually white. This kind of tube is called a **monochrome** tube. It is used in black and white televisions. The other kind of tube can produce many colours. This kind of tube is called **polychrome**. It is used in colour televisions.

Black and white

Here is a monochrome tube.

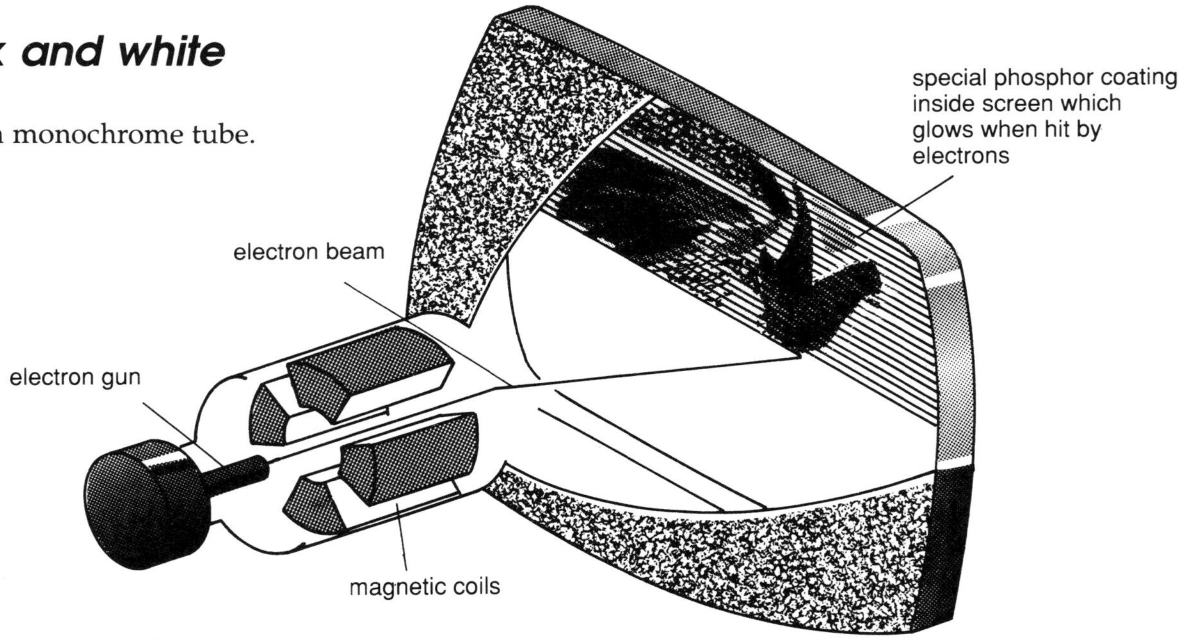

Nearly all the air is sucked out.

The **electron gun** fires a constant beam of electrons. Its a bit like a machine gun.

Magnetic coils move the beam of electrons up or down. Other coils move the electrons sideways. The electrons can be aimed at any point on the screen.

The **electron beam** is completely invisible. Light energy is produced only when the electrons hit the screen.

The **screen** is covered with dots of a special material. This material glows when it is hit by electrons. The more electrons hit a dot, the brighter it is.

The electron beam crosses the screen in straight lines. Your television picture is made up of 625 lines. Each line contains around 800 dots.

The picture is made of thousands of tiny dots. Some are bright. Some are dark. If you get very close to a black and white television you can see the dots. Do not do this for a long time. Your eyes may get sore.

When the pattern of bright and dark changes it looks as though the picture is moving.

Colour

Here is a polychrome tube.

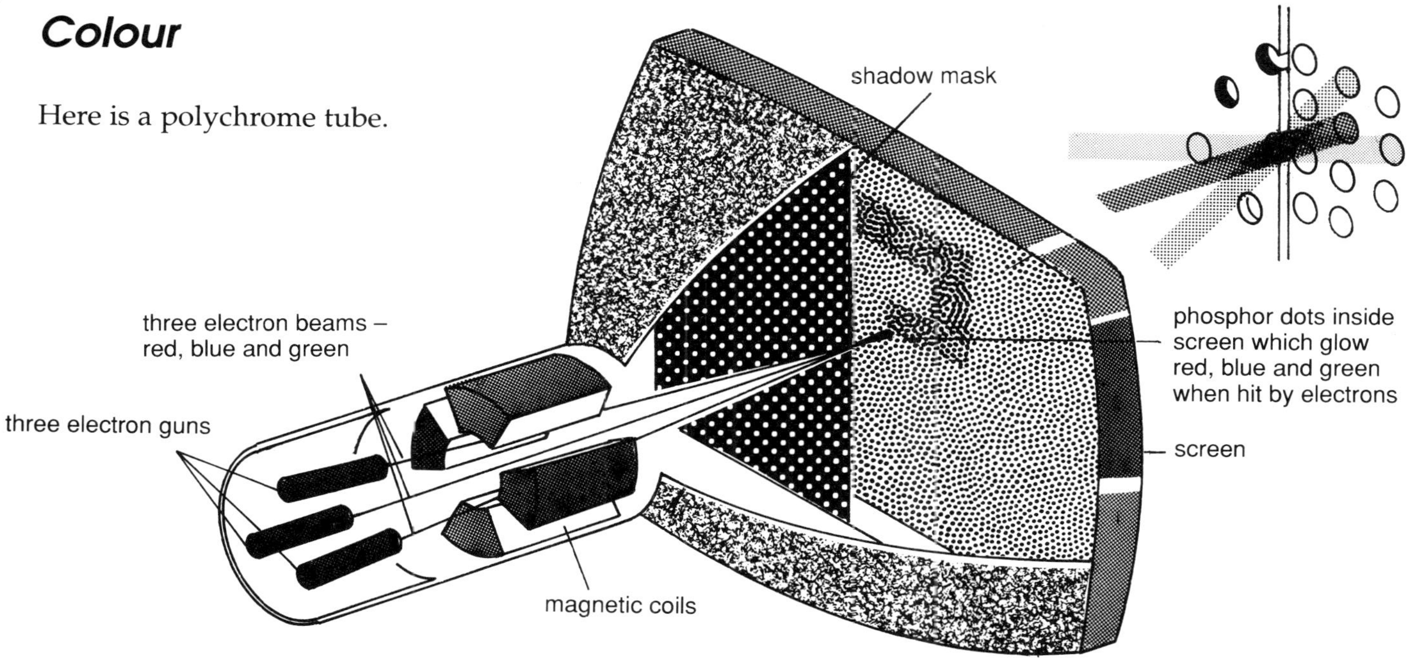

Nearly all the air is sucked out.

Three **electron guns** fire *three* separate beams of electrons. One beam is used to produce red light. Another is used to produce blue light. The third is used to produce green light.

Magnetic coils move the beams of electrons up or down. Other coils move the electrons sideways. All the beams are moved at the same time. The electrons can be aimed at any point on the screen.

The electron beams are completely invisible. Light energy is produced when the electrons hit the screen.

The **screen** is covered with dots. These dots are made of *three* special materials which glow when they are hit by electrons. One material glows red. Another glows blue. The third glows green. The dots are arranged in sets of three. Each set has one red dot, one green dot and one blue dot. Each dot is hit by a different beam of electrons.

The **shadow mask** makes sure that each beam of electrons can hit only one colour of dot.

The electron beams cross the screen in straight lines. A colour television screen is made up of 625 lines. Each line has around 800 sets of dots.

The picture is made of thousands of tiny dots. Some are bright. Some are dark. Different colours are made by changing the brightness of the red, blue and green dots. If you get very close to a colour television you can see the dots.

When the pattern of colours changes it looks as though the picture is moving.

Now try these!

C

1 a What does monochrome mean?
  b What does polychrome mean?

2 Why is most of the air sucked out of a television tube?

3 Would you be able to see a beam of electrons?

4 Why does a television tube have two sets of magnetic coils?

5 A television picture is made up of thousands of dots. How is the brightness of a dot changed?

6 How many dots are there in a black and white television picture?

7 Does a television picture have kinetic energy?

8 What three colours are used to make a colour television picture?

9 Why does a colour television need three electron guns?

10 Why does a colour television have a shadow mask?

11 How many dots are there in a colour television picture?

E

12 Name any other piece of electrical equipment which has a television tube.

13 What other colours of monochrome tubes are available, apart from white?

14 Find out about phosphors. Why are they important for televisions?

15 a What are the primary colours in painting?
   b Why are these colours not used for colour televisions?

light fantastic

Since earliest times people have used light energy to send messages. Modern telecommunications engineers still use light but now the light is put inside specially made fibres called **optical fibres**.

A fibre is any very thin strand of material. But optical fibres must be very clear. Ordinary glass, like the glass in your windows, is absolutely useless! If you ever get a chance to look at a pane of glass end-on you will be able to see why. The special glass used for optical fibres is so clear that you would be able to see right through a block several kilometres thick.

All optical fibres are very thin, even thinner than a strand of your hair. Usually fibres are only about 50 millionths of a metre thick. Despite this, they are all made with more than one kind of glass.

There are three different kinds of optical fibre. These are called **monomode, step-index multimode** and **graded-index multimode.**

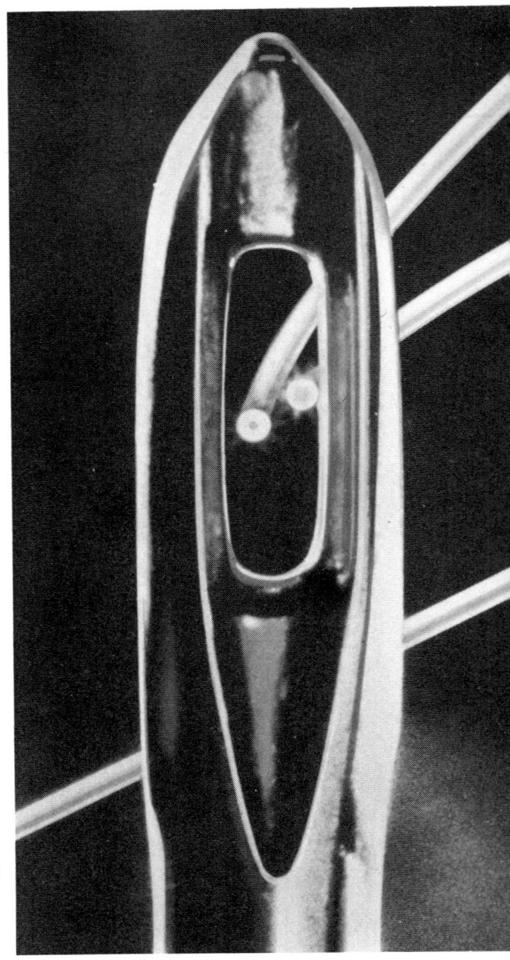

An optical fibre is thin enough to pass through the eye of a needle

The simple fibre

Monomode fibres are made from two kinds of glass. The centre of the fibre is made from one kind of glass and the outside is made from the other kind of glass. The central strand is only 5 millionths of a metre across.

Light signals are sent through the bit of glass in the centre. Once the light is inside this strand it cannot get out. It has to travel all the way to the end of the fibre. This is because of **total internal reflection**.

Total internal reflection occurs because light in the central strand travels more slowly than it would in the outside piece of glass.

A monomode fibre can carry only a single beam of light.

A monomode optical fibre

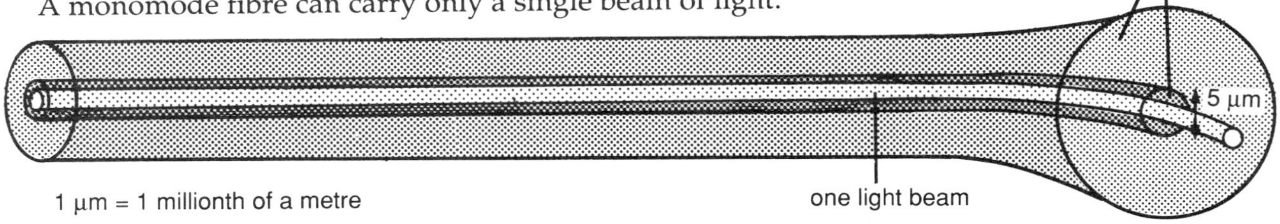

A better fibre

A **step-index multimode** fibre is made in the same way as a monomode fibre. But the central strand of the fibre is about ten times thicker.

This kind of fibre can carry lots of light beams at the same time. These light beams travel at different angles. But light beams at shallow angles get to the end of the fibre more quickly than light beams at steep angles. So the light in the fibre spreads out.

A step-index multimode fibre

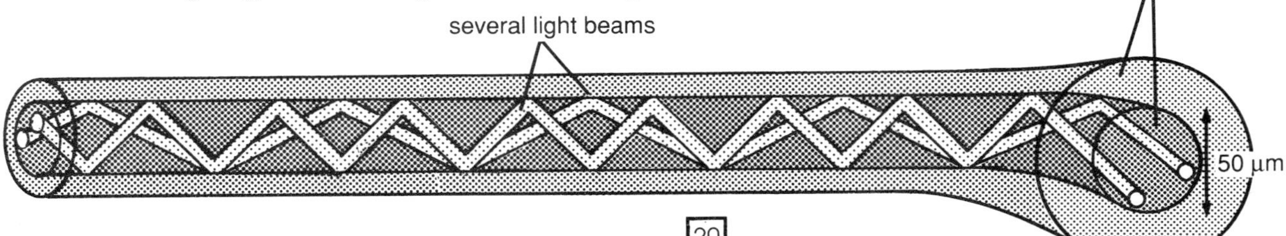

An even better fibre

A **graded-index multimode** fibre is the most difficult to make. It is also the most expensive. But it is the best available at the moment.

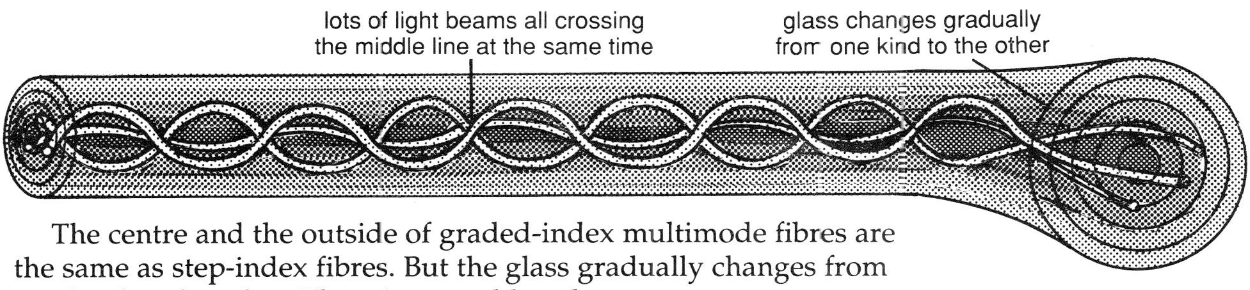

A graded-index multimode fibre

The centre and the outside of graded-index multimode fibres are the same as step-index fibres. But the glass gradually changes from one kind to the other. There is no sudden change.

This kind of fibre can also carry lots of light beams simultaneously. Notice that the beams of light all cross the middle line at the same point. The light in the fibre doesn't spread out any more.

Also the light is not being reflected. It turns because its speed gradually changes as it moves across the fibre.
The light is refracted.

Good points and bad points

In some ways optical fibres are very good. For example, a single optical fibre can carry about 100 000 times more information than a wire. And they can be laid under water quite safely as the fibres are not electrical. This also means that they are not affected by electrical interference. Lastly, optical fibres waste less energy than electrical energy in wires. Signals need to be boosted less often when information is being send a long distance.

But light energy in a optical fibre travels slower than electrical energy in a wire. Inside a fibre the light travels at around only 200 million metres per second. Slow, huh!

Seeing the light

To make the best use of optical fibres scientists have had to develop special tiny electronic transducers. At the start of a fibre the transducer changes electrical energy to light energy. At the end the transducer changes the light energy back to electrical energy.

FASCINATING FACT!

You could not see the light used in optical fibres. It is invisible. The light used is actually **infrared light**. Sometimes infrared light is called **heat rays**.

C — NOW TRY THESE!

1 What is a fibre?

2 How many kinds of optical fibre are there?

3 Why are optical fibres not made from ordinary glass?

4 How could you tell the difference between a monomode fibre and a step-index multimode fibre?

5 What is the difference between monomode and multimode fibres?

6 Which type of optical fibre is best? Give reasons for your choice.

7 What kind of fibre uses refraction?

8 Are optical fibres better than wires for carrying messages? Explain your answer.

9 Write down the energy change at the start of an optical fibre.

10 Write down the energy change at the end of an optical fibre.

e

11 a Why are telecommunication cables sometimes laid under water?
b Why are optical fibres suitable for underwater cables?

12 What is electrical interference?

13 Why does energy need to be boosted for telecommunication over long distances?

14 Why is heat important to optical telecommunication?

15 a Find out about total internal reflection.
b Why is it called total?
c Explain why refraction is impossible when total internal reflection occurs.

SPECIAL AERIALS

An **aerial** is used for sending out or for receiving signals. These signals are usually carried by radio waves. Modern telecommunication systems use special aerials to send messages over very large distances. These aerials are called **dish aerials** and look like big saucers. Dish aerials use the reflection of waves.

When a signal hits the dish it bounces off. So though the energy starts out in the wrong direction it is soon going in the right direction. Also, the energy coming from the dish is shaped into a beam.

This beam of energy hardly spreads out at all. It travels in one particular direction. So the energy can be sent much bigger distances.

The wrong way!

When an ordinary aerial sends out a signal, the signal travels in all directions. So the energy of the signal is spread out thinner the further it goes.

A dish aerial is different. It does not send out the energy in all directions. It does not even send the energy in the direction it wants the message to travel. It send the energy straight at the dish! Aren't scientists strange!

Two-way traffic

Also, if a signal has come a very large distance it might be too weak for an ordinary aerial. A dish aerial can help here too. All of the energy which hits the dish is bounced back to one point. Put a receiver at this point and its gets all the energy.

This receiver must be facing the dish. That is, it should have its back to the sender of the signal.

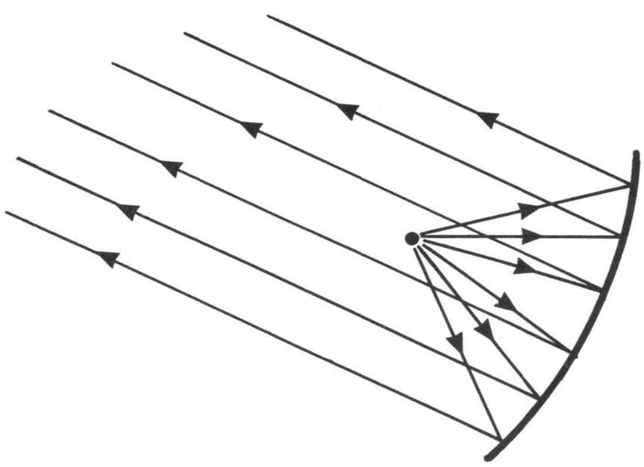

A dish aerial sending out a beam of energy

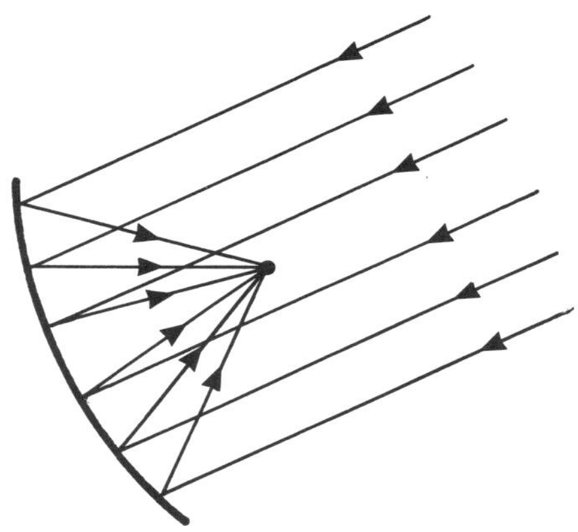

A dish aerial receiving a beam of energy

Take aim

Because they send signals in one particular direction, dish aerials have to be aimed very carefully. Usually, a dish aerial is pointed straight at another dish aerial. Both can then be used for sending and receiving messages. If either aerial is pointing in the wrong direction, it is impossible to send messages between them. The bigger the distance between two dish aerials the more difficult it is to get the aim just right.

This is not a problem with ordinary aerials. They send messages in all directions. And they can pick up messages from any direction.

Dish aerials have to point straight at each other to send messages to each other

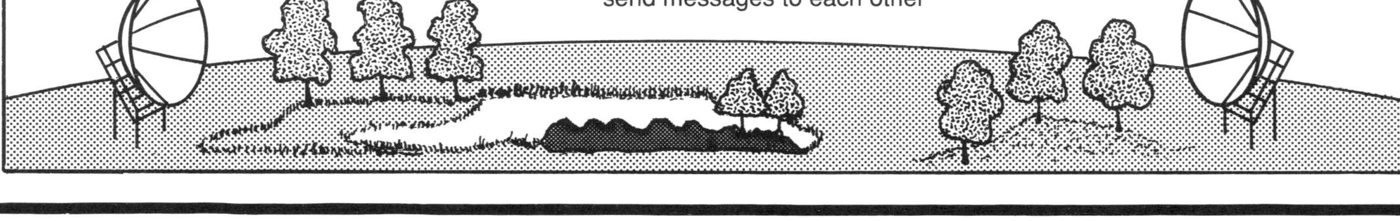

Telephone microwaves

The modern telephone system uses lots of dish aerials. Look at the top of the British Telecom tower. See the small dish aerials. These aerials are used for sending telephone messages. If you travel about the country you might see some small pylons with dish aerials. These are links in a network of telephone towers which cover almost the whole country.

This network of telephone towers uses a kind of microwave to carry the messages. These microwaves are not the same as the microwaves used in cookers. So there is no danger of you being cooked if you get too close.

Dish aerials at the top of the British Telecom tower

Now try these!

C

1 Why are special telecommunication aerials called dish aerials?

2 What form of energy is usually used with dish aerials?

3 Which wave effect do dish aerials use?

4 Give one difference between a dish aerial and an ordinary aerial.

5 **a** Explain why a signal from an ordinary aerial gets weaker as it gets further from the aerial.
b Does this happen to a signal from a dish aerial? Explain.

6 Why does the receiver of a dish aerial have its back to the sender of the messages it is trying to pick up?

7 **a** Why do dish aerials have to be aimed very carefully?
b Why is aiming ordinary aerials not so important?

8 What is a network?

9 Some people now have dish aerials in their gardens. Why?

e

10 Why is it more difficult to aim dish aerials which are far apart?

11 **a** Find out what a parabola is.
b What is the connection between a parabola and dish aerials?

12 Should dish aerials be made from metal, plastic or wood? Explain your answer.

13 Astronomers use very big dish aerials to 'look' at the sky.
a Why is the word 'look' not quite right here?
b What form of energy are the astronomers detecting using these aerials?
c Why do astronomers use this kind of energy?

Angels of Telecommunication

The word **satellite** can be used to describe any object which moves around another object. For example, the planet earth is a satellite of the sun and the moon is a satellite of the earth. In modern telecommunications, a satellite is usually a specially built radio or television station put into space. These are the 'angels of telecommunication'.

Telecommunication satellites move around the earth. One complete trip all the way round is called an **orbit**. If a satellite is going at just the right speed, it stays at the same height above the planet. As a general rule, the higher the satellite, the slower it should travel to maintain its height.

Do satellites fall?

The answer is yes! Satellites are pulled down by the force of gravity so they do fall. But the surface of the planet is not flat. It is curved.

Suppose the satellite starts at A. Moving to B it falls a distance d. At A it was a height h above the planet. At B it is still a height h above the planet. The satellite fell but it did not get any closer!

If a satellite is going too slow it gets lower. This helps speed it up. If it is going too fast, it gets higher. This helps slow it down.

Satellites fall but do not get any closer

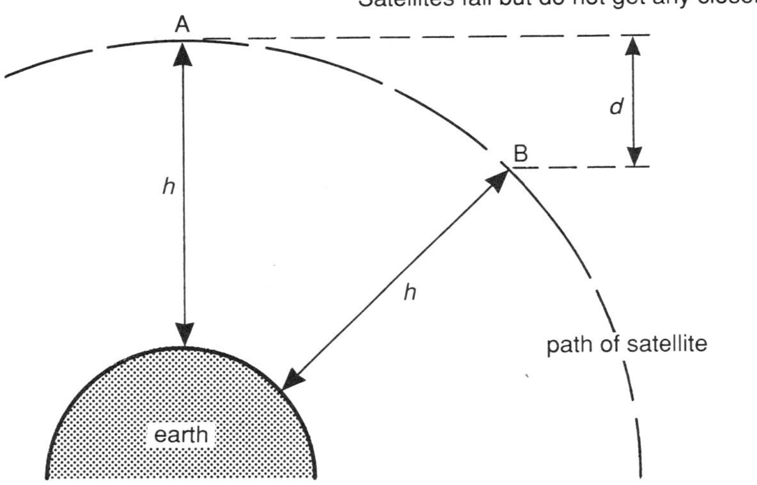

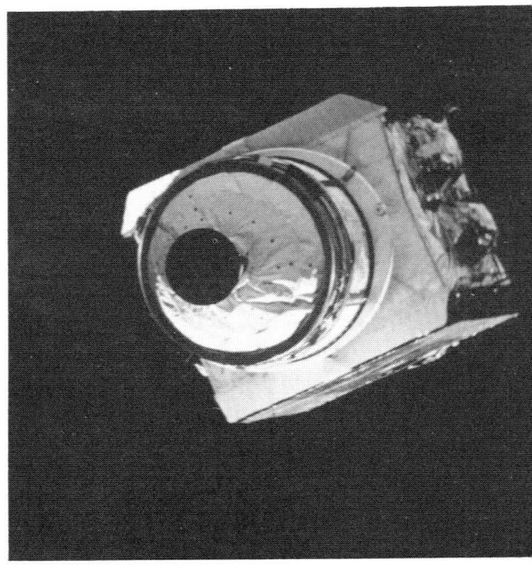

A telecommunications satellite in space

Fiction becomes fact

In 1945, Arthur C. Clarke, scientist and science fiction writer, suggested that a satellite 36 000 kilometres above the equator would take 24 hours to complete an orbit. He said that a satellite going in the same direction as the spinning planet would stay above the same point on the surface. From the earth, he said, this satellite would look as if it was not moving. It would be **geostationary**. When Clarke said all this, no one had ever put a satellite of any kind into space. Many people thought this was science fiction! Only nineteen years later, in 1964, SYNCOM, the first experimental geostationary satellite, was launched into a 36 000 kilometre orbit. Clarke was right! The satellite did stay above the same point on the surface. The satellite did look as if it was not moving!

Into space and back again

To communicate with satellites, special stations must be built on earth. At these stations a message is converted into a radio signal. Using a dish aerial this signal is sent to the satellite. But the aerial must be pointing straight at the satellite. If the direction is out by even one-tenth of one degree the energy reaching the satellite is cut by half. The satellite picks up the radio signal and boosts it to make it stronger. The signal is then sent to a different earth station. To avoid confusion satellites receive signals and send signals using different frequencies of radio waves.

Communication between continents

With just three geostationary satellites in orbit it is possible to send messages anywhere on the earth. These messages can be sent very quickly as radio waves travel at the speed of light. Nowadays satellites are used to send telephone calls, radio programmes, letters and television pictures all over the globe.

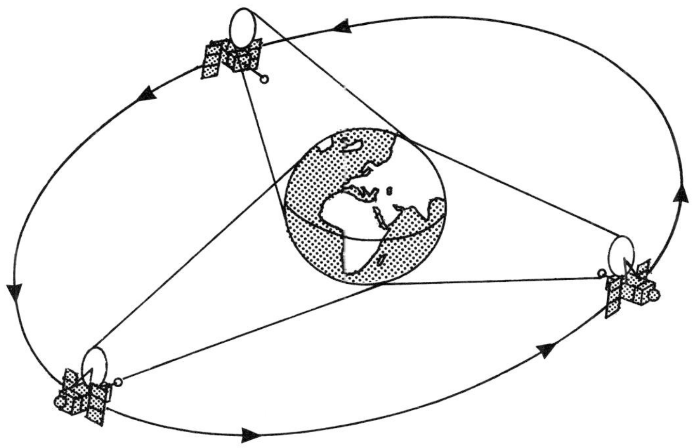

Three satellites can send messages all over the earth

FASCINATING FACT!

A satellite skimming the surface of the earth would have to fly at nearly 8000 m/s!

The cheapies?

Most telecommunication satellites are much lower than 36 000 kilometres. They are not geostationary. They keep moving over the surface of the earth so they can only be contacted every so often. However, it is much cheaper to put a satellite into a lower orbit. But it still costs millions of pounds so don't get the idea that they really are cheap!

NOW TRY THESE!

c

1. Explain in your own words what a satellite is.
2. What is an orbit?
3. How do you know that the force of gravity acts on satellites?
4. What does geostationary mean?
5. The speed of radio waves is 300 million m/s. Calculate the time for a radio wave to travel from earth to a geostationary satellite.
6. Why does a geostationary satellite have to take 24 hours for one orbit?
7. Geostationary satellites are much higher than other satellites. This makes them more expensive to launch. Do you think this is a waste of money? Explain.

e

8. In a distant galaxy, the planet Darth is the same size and shape as our own planet earth. But Darth does not spin.
   a Would it be possible to put a satellite into 'Darthstationary' orbit?
   b Explain your answer.
9. Not all satellites are used for telecommunications. Write down other possible uses for satellites.
10. Try to repeat Arthur C. Clarke's calculation. As well as the usual speed=distance/time equation you will also need the following information

$$(\text{speed in orbit})^2 = (G \times M) / R$$

where G = Newton's gravitation constant $= 6.7 \times 10^{-11}\ \text{N m}^2\ \text{kg}^{-2}$
M = mass of earth $= 6.0 \times 10^{24}$ kg
R = radius of the orbit in metres

You need the radius of planet earth too! It is
r = radius of the earth $= 6.4 \times 10^{6}$ m

11. Explain why a satellite should get faster if it falls closer to the earth.

COMPUTERS TALKING

Computers in the same room can pass messages backwards and forwards very easily. The computers are connected directly to each other. But computers which are far apart cannot be linked like this – they have to use the phone!

No problem?

Ordinary telephone lines are a bit of a problem .

Ordinary telephone lines carry weak electric currents. When someone speaks into a telephone the current changes smoothly. A smoothly changing current is called an **analogue signal**. Telephone lines are very good at carrying analogue signals.

Computers send out short pulses of current. The current has only two values, on and off. These currents switch very quickly. A sharply changing current is called a **digital signal**. Telephone lines are very poor at carrying digital signals.

Any answers?

One bright spark thought of changing the digital signals into sound. The sounds can then be sent through the telephone line. At the other end the sounds can be changed back into digital signals. To do this, extra energy changers are needed. For computers the energy changers are called **modems**. Modem is short for **mod**ulator–**dem**odulator.

The computer's mouth and ears!

When you speak your mouth changes brain signals into sounds. The modulator part of the modem does the same job for the computer. When the computer is sending a message, the modulator makes bleeps, one tone for on and a different tone for off. Put the modulator at a telephone mouthpiece and the bleeps can be sent by telephone.

When you listen your ears change sounds into brain signals. The demodulator part of the modem does the same job for the computer. When the computer is receiving a message, the demodulator listens to the bleeps. The demodulator turns the bleeps into on and off signals. Put the demodulator at a telephone earpiece and the computer can listen to a message coming from another computer.

Different voices

Anyone talking on the telephone can tell the difference between his or her own voice and the voice of the other person. Computers need to be able to do this too. So different modems use different tones. When a computer is sending and receiving messages it listens to the other tones and ignores its own. If all modems made the same bleeps computers would be terribly confused.

Some modern telephones use digital systems. They turn the digital signals to sound too. Maybe you have heard the bleeps when you are 'dialling' a number on a digital telephone. If not, ask your parents – they might have a telephone like this at work. Every number has a different tone.

Limits

Modems sometimes work very well but there are often a lot of problems. For example, British modems and American modems cannot talk to each other. They do not 'speak' the same 'language'.

Today's modems run at 300 **baud**. They can send 300 **bits** of information a second. Three hundred bits is the equivalent of 30 individual letters. It would take a modem over $1^1/2$ seconds to send this sentence.

Improvements

Some people are working on modems which will run at 1200 baud and 2400 baud. These may become available soon. British Telecom are even working on a modem which runs at 9600 baud. But this only works over short distances.

Communications scientists are looking for a way of getting rid of modems altogether. Computers will then be able to use the telephone line as easily as you!

c — NOW TRY THESE!

1 What are analogue signals?

2 What are digital signals?

3 What kind of signals do computers send out?

4 Why do computers have a problem sending messages through ordinary telephone lines?

5 What is modem short for?

6 Why does a modem use two different tones?

7 Why do different modems use different tones?

8 In a digital telephone, why does every number have a different tone?

9 Modems may soon become obsolete. What does this mean?

10 Estimate how long it would take a 300 baud modem to send all the text on this page.

e

11 Why is it impossible for a computer in Glasgow to communicate with a computer in Boston?

12 **a** The Hayes Protocol is a set of modem commands. It was adopted as a *standard* by nearly all modem manufacturers. Explain what this means.
b Why does having a standard make modems easier to use?

13 **a** Why are faster modems better than slower modems?
b Estimate how long it would take a 9600 baud modem to send this book.

14 Can you find a link between modems and a pioneer of the telegraph?

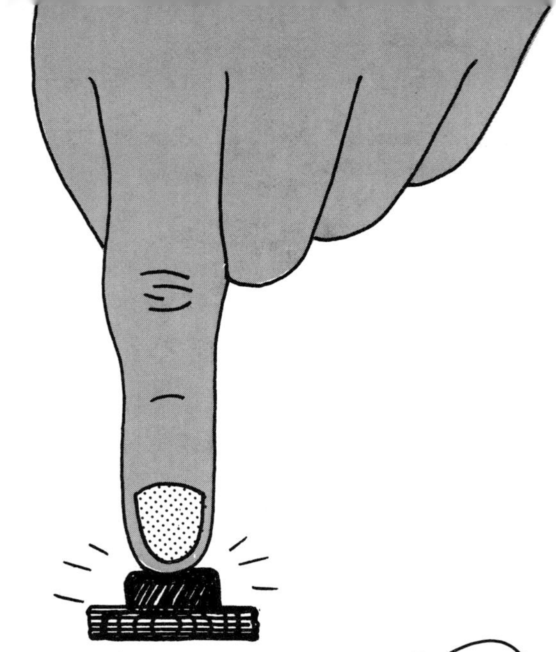

FINGER ON THE PULSE

At first modern telecommunications systems carried **analogue signals**. These signals change smoothly all the time. Nowadays more and more systems carry **digital signals**. Digital signals change very sharply. Messages are sent in pulses which switch off and on very rapidly. Usually the pulses are electrical or pulses of light.

At first you might think that pulses would not be very good for sending messages because all you can do is have the pulse on or off. But you would be wrong. Telecommunications scientists have come up with all sorts of ways of putting information into pulses. All these ways use very high frequency pulses to carry the information. But the information modulates the pulses in different ways.

a PAM is short for **pulse amplitude modulation**. Information is used to change the heights of the pulses. A PAM set of pulses would look like this:

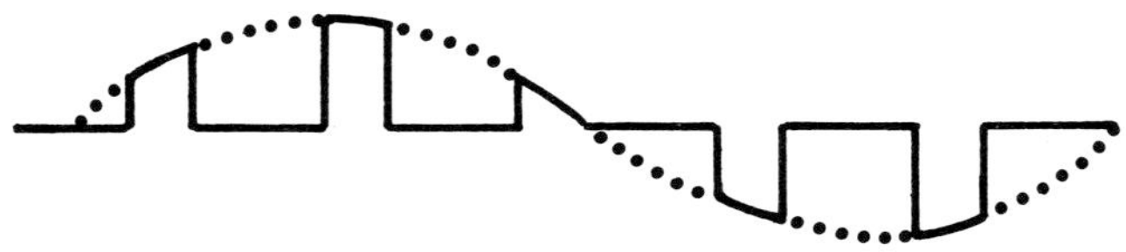

b PFM is short for **pulse frequency modulation**. Information is used to change the number of pulses in a second. A PFM set of pulses would look like this:

c PPM is short for **pulse position modulation**. This uses very narrow pulses. Information is used to move these pulses forwards or backwards in time. A PPM set of pulses would look like this:

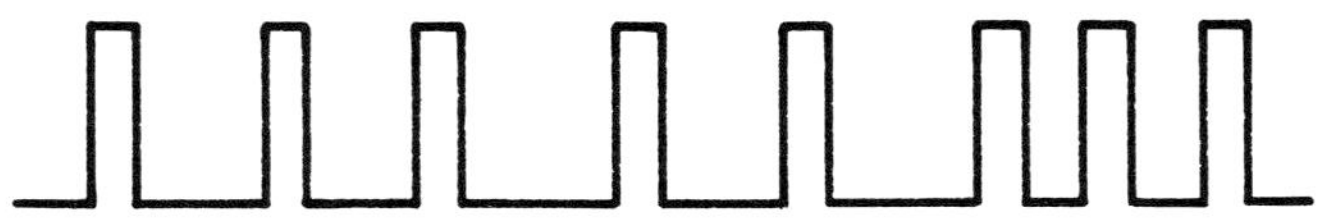

d PWM is short for **pulse width modulation**. Information is used to change the width of the pulses. A PWM set of pulses would look like this:

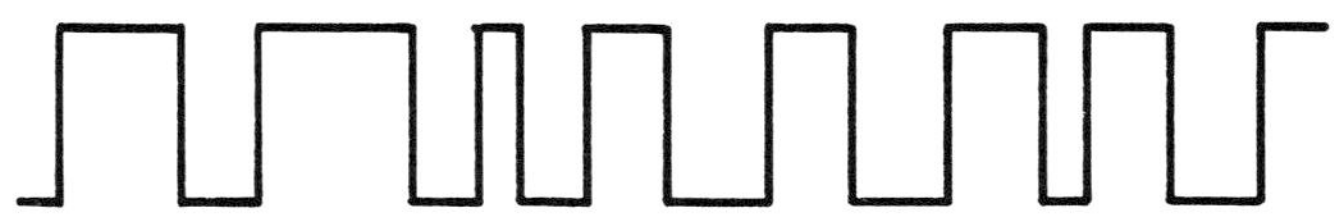

e PCM is short for **pulse code modulation**. It is the most useful type of pulse modulation. Digital transmission of television pictures, digital telephone systems and electronic keyboards all use PCM.

The possible range of signals is split into layers. Each layer is given a binary value. This is called its **code**.

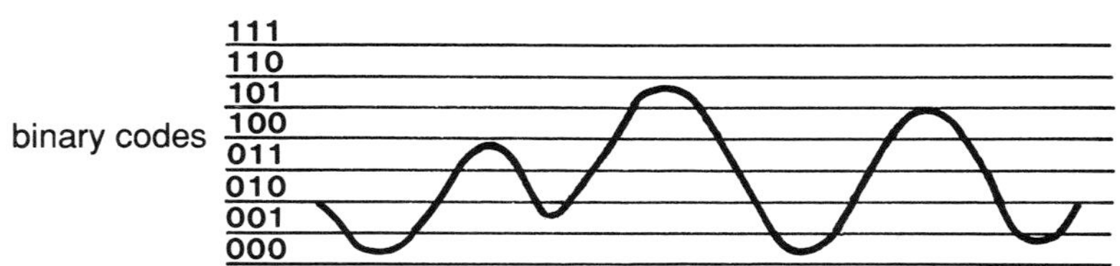

When a signal comes in, its value is measured every so often. Each measurement is given the code for its layer. So the signal is converted to a series of codes. These codes are what carry the information. A PCM set of pulses looks like this:

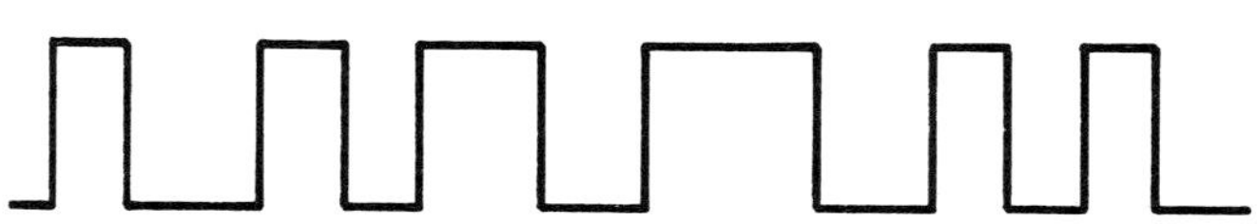

Two things are very important: the number of layers and the number of measurements made every second.

Mucking about

One of the fun things about PCM is that you can get special effects by changing the way the pulses are coded. Sometimes television producers do this deliberately to make your programmes more interesting.

The photographs below show how changing the number of layers can affect the picture. The picture gets clearer as more layers are used! All sorts of different colours can be added to the picture so that some really weird effects are produced. Lots of pop videos use this.

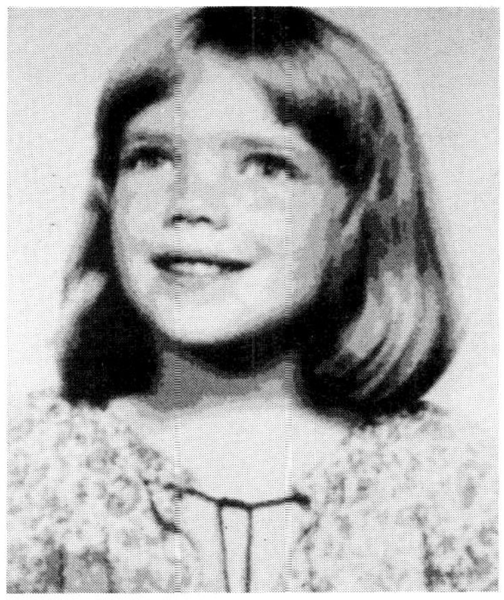

Also, whole blocks of a picture can be given the same code. If these blocks are very big then it is almost impossible to recognize the picture. Some television quiz shows use this idea when they are making up their questions.

Or the blocks can be different sizes in different parts of a picture. So one part of a picture can be very clear while another part is disguised. This is sometimes used to hide someone's face when they are being interviewed.

All of this is only possible with pulses. These effects cannot be used if an analogue system is used.

FASCINATING FACT!

In digital colour television systems the three colour signals are combined to give a black and white signal before it is transmitted. At the receiver the three colour signals are separated and then put together again to make the colour picture.

NOW TRY THESE!

1 What does modulation mean?
2 Do pulse modulation systems use high or low frequencies?
3 What forms of energy can pulses have?
4 How many different ways can pulses be used to send messages?
5 Explain PAM.
6 Draw a set of PAM pulses.
7 How is PWM different from PAM?
8 Which pulse modulation system uses very narrow pulses?
9 Which pulse modulation system is the most useful?
10 What is PCM short for?
11 What is a binary value?
12 Name two telecommunication systems which use PCM.
13 What are the two most important parts of PCM?
14 Do you think that pulses make telecommunication better? Discuss this with other members of your class.

15 DPCM is short for **differential** pulse code modulation. Find out how DPCM is different from PCM.
16 For the transmission of sound by PCM, all the layers have different depths. For picture transmission all the layers have the same depth. Why do you think there should be this difference?

Modern Telephones

The modern telephone system can do all sorts of things which used to be impossible. One reason is that the system is changing from an analogue system to a digital system. Here are just a few of the things which might amaze your granny, and maybe your dad as well.

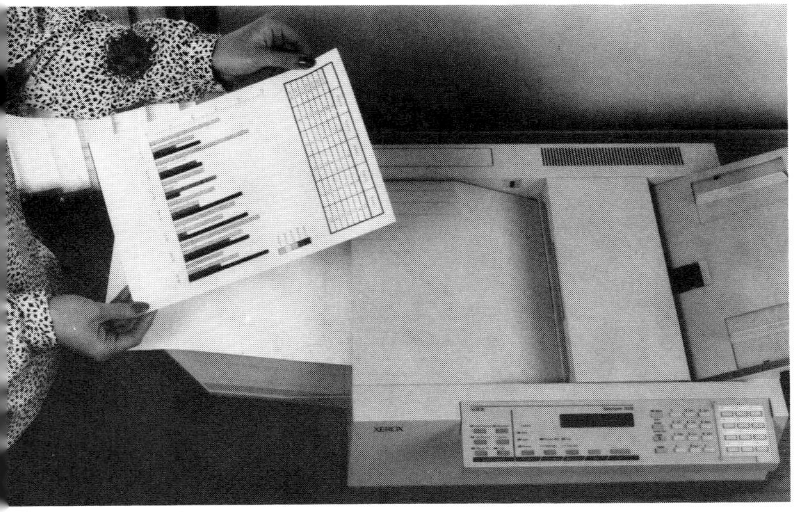

A fax machine

The facts about fax

The word fax is a badly spelt shortened form of the word **facsimile** (fak-sim-ill-eh). If you look this word up in a dictionary you will find that it means an exact copy.

A fax machine uses the telephone system to send an exact copy of any document from one place to another. Usually letters are sent this way, but photographs can be sent too.

The sending fax has a scanner which crosses a document in a series of parallel lines. It measures how bright or how dark each part of the line is. This information is converted to a stream of pulses which are then sent through the telephone lines. At the other end, the receiving fax converts the stream of pulses back to a pattern of bright and dark. This pattern is then printed on a blank sheet of paper.

The fax makes an **electronic mail** system possible. People do not have to send letters through the post. They can send them through the telephone lines!

Instant banking – the good news

Home computers can be linked to bank computers through the telephone lines. Customers with the right computer program can ask the bank all about their own bank account any time, day or night. They can also find out what kind of services the bank is offering and how much interest it will charge.

Instant banking – the bad news

But some big stores have tills which are connected through the telephone lines to banks.

When you want to buy something the till can ask the bank's computer if you have enough money. If the answer is yes, the till will then tell the bank to transfer the money from your bank account to the store's bank account. This happens right away, while you are still standing at the counter. Instant poverty!

Even worse, if the bank's computer tells the till you don't have enough money then you probably won't get your groceries. Instant hunger!

Prestel

The Prestel system combines a special telephone, a television and a big computer **database**. The special telephone is really a mixture of a telephone and a home computer. The computer database is like a big information store held on a mainframe computer. But any television will do.

You use the telephone to ask the database for information. When information is sent back to you it can be displayed on the television screen, stored in the home computer memory or printed out on paper.

But the system can refuse to send you information. If you try to poke your nose into someone else's business the big computer won't let you! Everyone using the system can have privacy.

This system is a bit like teletext. But teletext is only one way. Prestel is a two-way system.

More inventions

This unit is not a complete list of all the new telephone inventions. For example, we have not mentioned telephone answering machines or radiotelephones. Perhaps you can think of some others! But it might give you an idea of how much telephone communication has changed since the days of Bell (and Gray!).

A radiotelephone doesn't need wires

Now try these!

c

1 What is a fax?

2 **a** Do you think a fax system uses analogue or digital signals?
b Explain your choice.

3 Work out all the energy changes in a fax system.

4 **a** Why does a fax scanner work in parallel lines?
b Should a fax printer work in parallel lines too?

5 Do you think a fax machine could be used to send colour photographs?

6 Do you think that instant banking is a good idea?

7 What is a mainframe computer?

8 Which is better, an ordinary telephone or a radiotelephone? Explain your answer.

e

9 **a** In what ways is the Prestel system better than teletext?
b In what ways is the teletext system better?

10 Find out about radiotelephone systems. Write a one-page note.

11 **a** Describe how a telephone answering machine works.
b Who might find a telephone answering machine useful?

12 Write down all the energy changes between speaking into one radiotelephone and listening to another radiotelephone.

13 **a** Find out about System X. (Try writing to British Telecom.)
b Why are British Telecom changing to this system?

THE FUTURE

This unit is all about imagination. The only thing certain about the future of telecommunications is that no one really knows. A lot of the things which will happen haven't been discovered yet. Who knows? Maybe you will have a wonderful idea and become world famous! Here are some ideas to get your imagination started.

Telephone

One big problem with telephones is that they are not very useful for deaf people. Videophones could help. These are telephones which send pictures as well as sound. You can see and hear whoever you are talking to. Deaf people could then communicate using sign language!

Also, suppose a telephone could translate spoken words into print. A videophone could then print out words on its screen. If it had a keyboard, messages could be typed out. Deaf people could communicate with anyone.

Have you ever telephoned another country where the people don't speak English? It can be very difficult. This is one good reason for learning other languages. But perhaps, one day, telephones will be able to translate from one language to another. People in different countries will be able to communicate easily!

Radio

Have you ever had trouble tuning in to a particular radio station? Someday you might be able to press a button and the radio will find the station you want. This would not be like a car radio which is pre-tuned for five or six stations. It would work for any radio station!

Many people have portable radios. They can listen to their favourite radio station everywhere they go. Some even have portable radio intercoms. They can send and receive messages over short distances. But you could have a portable satellite radio set. You could send and receive messages anywhere in the world!

Television

Some scientists are working on making television sets smaller and lighter. Already, small televisions are available, but can you imagine a television set as small as a pack of cards? That really would be portable!

Many televisions now have remote controls. But someday you might be able to tell the television what to do. It would be a **voice-activated** television. You say 'switch on' and the television goes on, 'switch off' and it goes off again. You could change channels, turn up the volume...everything you can do with a television now. It could even be made so that it only works for your voice! This might make some people very lazy, but would be very useful to people who have trouble moving around.